SCIENTIFIC AMERICAN EDUCATIONAL PUBLISHING

SCIENTIFIC AMERICAN INVESTIGATES SPIDERS

TRAP-DOOR SPIDERS

BY ROXANNE ALDOLPHO

Published in 2025 by The Rosen Publishing Group
in association with Scientific American Educational Publishing
2544 Clinton Street, Buffalo NY 14224

Library of Congress Cataloging-in-Publication Data
Names: Aldolpho, Roxanne, author.
Title: Trap-door spiders / Roxanne Aldolpho.
Description: New York, New York : Scientific American Educational Publishing, an imprint of Rosen Publishing, [2025] | Series: Scientific American investigates spiders | Includes bibliographical references and index.
Identifiers: LCCN 2024006018 | ISBN 9781725351332 (library binding) | ISBN 9781725351325 (paperback) | ISBN 9781725351349 (ebook)
Subjects: LCSH: Trap-door spiders–Juvenile literature. | Spiders–Juvenile literature.
Classification: LCC QL458.4 .A383 2025 | DDC 595.4/4–dc23/eng/20240319
LC record available at https://lccn.loc.gov/2024006018

Portions of this work were originally authored by M. H. Seeley and published as *Trap-Door Spiders*. All new material in this edition is authored by Roxanne Aldolpho.

Designer: Andrea Davison-Bartolotta
Editor: Brianna Propis

Photo credits: Cover, p. 1 Ken Griffiths/Shutterstock.com; p. 5 zaidi razak/Shutterstock.com; pp. 7, 9, 11, 21 Pong Wira/Shutterstock.com; p. 13 (inset) Sixplus/Shutterstock.com; p. 13 (main) Dave Pinson/Alamy Stock Photo; p. 15 (funnel-web) James van den Broek/Shutterstock.com; p. 15 (trap door) Young Swee Ming/Shutterstock.com; pp. 17, 19 (inset) Ken Griffiths/Shutterstock.com; p. 19 (main) Blue Pebble/Shutterstock.com.

Printed in the United States of America

CPSIA compliance information: Batch #CSSA25. For Further Information contact Rosen Publishing at 1-800-237-9932.

CONTENTS

Words in the glossary appear in **bold** type the first time they are used in the text.

UNLIKE OTHERS

If you stumbled upon a spiderweb in a corner of your house, it wouldn't belong to a trap-door spider. In fact, you won't ever see a trap-door spider in your house at all. Trap-door spiders don't make webs—and they don't hang from tree branches, either!

These creatures live in **burrows** underground, usually hidden behind a door covered in dirt and plant matter. The spider produces silk that it uses to hang a door on its home. Silk lines are spread out around the burrow's entrance, so the spider has an alarm system to alert it of any visitors!

Blink and you may miss it! This trap-door spider is hiding in its underground burrow.

FUN FACT

TRAP-DOOR SPIDERS ARE EXCELLENT AT **CAMOUFLAGING** THEIR HOMES. IT IS ALMOST IMPOSSIBLE TO SPOT THEIR BURROW UNLESS THE DOOR IS OPEN!

SIMILAR SPECIES

Trap-door spiders can be found in many families and species, or kinds, of spiders. In the *Liphistius* group alone there are 50 species of trap-door spider. Each has special habits and ways of building their burrow. Some types of trap-door spiders are more common than others, but most are found where it's warm.

Though each species looks and acts somewhat differently, they all use a trapdoor to catch **prey** and hide from predators. These species are mostly nocturnal, meaning more active at night, and catch food by hiding in their burrow until prey passes by.

A TRAP-DOOR SPIDER'S BURROW

Some burrows are several feet underground, while others are only 1 inch (2.5 cm) deep. Sometimes trap-door spiders build a second door that is also made from silk.

FUN FACT

A SPECIES OF TRAP-DOOR SPIDER DISCOVERED IN 2012 WAS NAMED AFTER FORMER PRESIDENT BARACK OBAMA. IT IS CALLED THE *APTOSTICHUS BARACKOBAMAI.*

NOCTURNAL HUNTING

During the day, trap-door spiders sleep. This means they have lots of energy to hunt at night! They lay lines of silk out from their door and wait for something to wander by. The silk **vibrates** as it's walked on, and the vibrations help the spider figure out how big and how close the creature is.

If they sense the creature is the perfect size for a meal, trap-door spiders pop out of the door, bite the prey, and bring it back into the burrow to eat! Trap-door spiders eat a variety of things, such as grasshoppers, crickets, frogs, baby birds, baby snakes, mice, and even small fish.

A trap-door spider will wait patiently in its burrow until it senses prey is nearby. If it doesn't like the prey it catches, the spider will toss the prey out of the door alive.

FUN FACT

LIKE MOST SPIDERS, TRAP-DOOR SPIDERS MAY GO WEEKS—OR EVEN MONTHS—WITHOUT EATING!

DINNER GUEST

The average life span of a trap-door spider is 20 years. Females generally live longer than males. The males must leave their burrow to find a female to make babies with, which may be difficult because all the females are in hiding!

When a male finds a burrow, he must be very careful when he goes knocking—if the female thinks he's a predator, she won't let him in. And if she thinks he's prey, she might eat him. Even if she doesn't think he's prey, she may eat him anyway!

Spiderlings, or baby trap-door spiders, live with their mother for a few months before leaving her burrow to dig their own.

FUN FACT

A MALE TRAP-DOOR SPIDER MAY ALSO BE KILLED BY A FEMALE AFTER MATING IF HE CAN'T GET AWAY IN TIME.

HOME, WARM HOME

Since there are so many kinds of trap-door spiders, they are **native** to many places in the world. Usually, trap-door spiders like to live where it's warm, because the ground is soft enough to dig their burrows. They're often found in Australia, Asia, Africa, South America, and North America—especially the warmer parts of North America, like the Southeast and West.

Though these spiders may live in your area, you may never see one, because they're so often hiding in their burrow. Trap-door spiders are known for being shy!

Some trap-door spiders have doorless burrows! These are known as brown trap-door burrows and are often found scattered over lawns.

FUN FACT

THE GROUP OF TRAP-DOOR SPIDERS CALLED *ANAME*, OR WISHBONE SPIDERS, LIVE IN AUSTRALIA AND CREATE BRANCHED TUNNELS THAT HAVE EXITS TO THE SURFACE WITHIN THE BRANCHES.

NONAGGRESSIVE SPIDERS

Most trap-door spider species are nonaggressive, which means that they don't like to fight with other creatures. Though they look scary at 1/2 inch to 3 inches (1.5 to 3 cm) in size, most trap-door spiders just want to be left alone.

A bite from a trap-door spider is usually not very harmful to humans. The area around a trap-door spider bite will hurt a lot and probably swell up, but there won't be any lasting harm. Still, if a trap-door spider bites you, it's best to go see a doctor.

Trap-door spiders are often mistaken for funnel-web spiders, but a trap-door spider's bites are far less deadly.

FUN FACT

THE **VENOM** OF TRAP-DOOR SPIDERS MAY BE HELPFUL TO THE CREATION OF PESTICIDES, WHICH ARE MIXTURES OF CHEMICALS USED TO KILL BUGS.

CHEWING DIRT

The body of a trap-door spider is designed to help them survive. Their **jaws** are different from some other types of spiders because they have sharp bumps along their jaw that act like teeth. But trap-door spiders don't use these "teeth" for chewing food—they use them for chewing dirt when they're digging their burrow!

Trap-door spiders also have two large **fangs** that they use to capture their prey. Some species have extra mouthparts called pedipalps (PEH-duh-pahlps). Pedipalps help a trap-door spider dig, shape silk for burrow building, and eat.

Pedipalps are somewhat like a fifth pair of legs.

FUN FACT

A TRAP-DOOR SPIDER CAN HAVE FANGS AS LONG AS 1/2 INCH (1 CM)!

PREDATOR AND PREY

Trap-door spiders aren't picky eaters! This makes them helpful for people's gardens because they'll eat a lot of the bugs that can harm plants. But the world is a dangerous place for these spiders.

Scorpions and centipedes will eat trap-door spiders. The spiders are also prey for birds, raccoons, skunks, and an Australian animal called a bandicoot. It makes sense why trap-door spiders like to stay hidden in a burrow. No predator ever wants to become another animal's prey!

FUN FACT

A GROUP OF FLIES CALLED *MILICHIIDAE* OFTEN STEAL FOOD FROM TRAP-DOOR SPIDERS AND MAY EVEN FEED DIRECTLY FROM A SPIDER'S MOUTH!

Birds take trap-door spiders directly from their burrows, but the bandicoot and other animals will dig them out. Yikes!

SHARING IS CARING

If a creature's habitat is being destroyed, the creature itself is in danger of **extinction**. This is the case for trap-door spiders. A species known as the Kanthan Cave trap-door spider is on the **endangered** species list because its home is in a single cave in Malaysia. The cave is at risk of being harmed from human use.

Even though they might look creepy, spiders play an important role in keeping nature healthy and your garden free of pests. We share the planet with these helpful creatures, and we need to keep them safe!

Only skilled spider handlers should ever touch a trap-door spider. It's best for most people to leave them alone in their burrows.

FUN FACT

SOME PEOPLE KEEP A TRAP-DOOR SPIDER AS A PET, BUT THIS IS DANGEROUS BECAUSE THE SPIDER COULD MISTAKE THEIR HAND FOR PREY AND BITE IT!

GLOSSARY

burrow: A hole in the ground made by an animal, or to make a hole in the ground.

camouflage: Disguising one's body or habitat to hide it from others.

endangered: At risk of dying out.

extinction: When a species of something dies out completely.

fang: A hard, sharp-pointed body part a spider uses to put venom in its prey.

jaws: The body part that makes up the mouth.

mate: To come together to make babies.

native: Where someone or something is from.

prey: An animal hunted by other animals for food.

venom: Poison created by one animal to harm another.

vibrate: To move back and forth quickly.

FOR MORE INFORMATION

Books

Higgins, Melissa. *Splendid Spiders.* North Mankato, MN: Pebble, 2020.

Lundgren, Julie K. *Spiders.* New York, NY: Crabtree Publishing Company, 2022.

Websites

Britannica Kids: Trapdoor Spider

https://kids.britannica.com/students/article/trapdoor-spider/313879

Learn more fun facts about this fascinating and sneaky spider!

NewScientist

https://www.newscientist.com/article/2408692-tap-dancing-spider-rediscovered-after-disappearing-for-92-years/

A species of trap-door spider has recently been rediscovered after 92 years! Read more about it here.

INDEX